YOUR KNOWLEDGE HAS VALUE

- We will publish your bachelor's and
  master's thesis, essays and papers

- Your own eBook and book -
  sold worldwide in all relevant shops

- Earn money with each sale

Upload your text at www.GRIN.com
and publish for free

Ekta Prakash

Analysis and Experiments of Carbohydrate

GRIN Publishing

Bibliographic information published by the German National Library:

The German National Library lists this publication in the National Bibliography;
detailed bibliographic data are available on the Internet at http://dnb.dnb.de .

Imprint:

Copyright © 2014 GRIN Verlag GmbH
Print and binding: Books on Demand GmbH, Norderstedt Germany
ISBN: 978-3-656-84802-8

This book at GRIN:

http://www.grin.com/en/e-book/284516/analysis-and-experiments-of-carbohydrate

ANALYSIS OF CARBOHYDRATE

EXPERIMENTS OF CARBOHYDRATE

Dr. Ekta Prakash

D.Phil (Biochemistry)

Allahabad University

<table>
<tr><th>S.No</th><th>CHAPTER</th><th>Page</th></tr>
</table>

Chapter I

Carbohydrate

1.1 Introduction:

Carbohydrates are compounds of carbon, hydrogen and oxygen with hydrogen and oxygen generally occurring in the ratio of 2:1, hence named carbohydrate (hydrates of carbon). Chemically, these are the aldehyde and ketone derivatives of polyhydric alcohol of polyhydric alcohol (Having more than one 'OH' group). The compounds of this class could be represented by the general formula C_x (H_2O) y. For example, we can write glucose ($C_6H_{12}O_6$) as C_6 (H_2O) $_6$; cane sugar ($C_{12}H_{22}O_{11}$); and so on. For this, reason, French named them as carbohydrates (hydrates of carbon). However; the name carbohydrate is a misnomer. While the inorganic hydrates contain loosely combined water, the compounds of this class contain no water molecules as such. They are, in fact, polyhydroxy aldehydes or ketones and it is a sheer coincidence that H and O are present in the ratio of 2 to 1. Now there is carbohydrate (rahmnose) ($C_6H_{12}O_5$) known in which in which the ratio of H and O is different. And then, it is also noteworthy that compounds such as formaldehyde (CH_2O) and lactic acid ($C_3H_6O_3$) which even though conforming to the formula $C_X(H_2O)_Y$ are not carbohydrates.

1.1.1 Classification of Carbohydrate:

Carbohydrates are grouped into three categories Monosaccharide, Disaccharides and polyssacharrides. The first two are commonly known as sugars and readily soluble in water while the later known as starch.

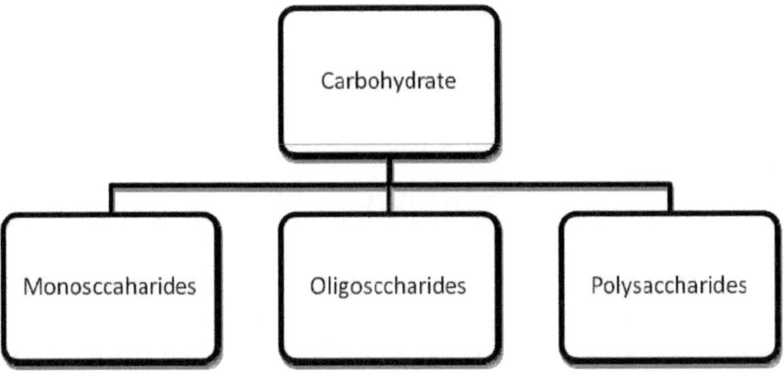

Fig1.Classification of Carbohydrate

1.1.2 Monosaccharide:

The simplest of carbohydrates are the monosaccharide that cannot be hydrolyzed e.g. glucose and fructose. Monosaccharides are either **aldehydes or ketone** with one or more hydroxyl groups; the six carbon monosaccharide glucose and fructose have five hydroxyl groups. The carbon atom to which hydroxyl group attached is often chiral centres or stereoisomerism is common among monosaccharide. Monosaccharides are colorless, crystalline solids that are freely **soluble in water** but insoluble in nonpolar solvents. The backbone of monosaccharide is an **unbranched** carbon chains in which all the carbon atoms are linked by **single bonds.** One of the carbon atoms is double-bonded to an oxygen has a hydroxyl group. If the carbonyl group is at an end of the carbon chain, monosccharide is an **aldehyde** and is called an **aldose**; if the carbonyl group is at any other position, the monosaccharide is a **ketone** and is called a **ketose.** The simplest monosaccharide is the two three carbon trioses; glyceraldehydes, an aldose, and dihydroxy acetone, ketose.

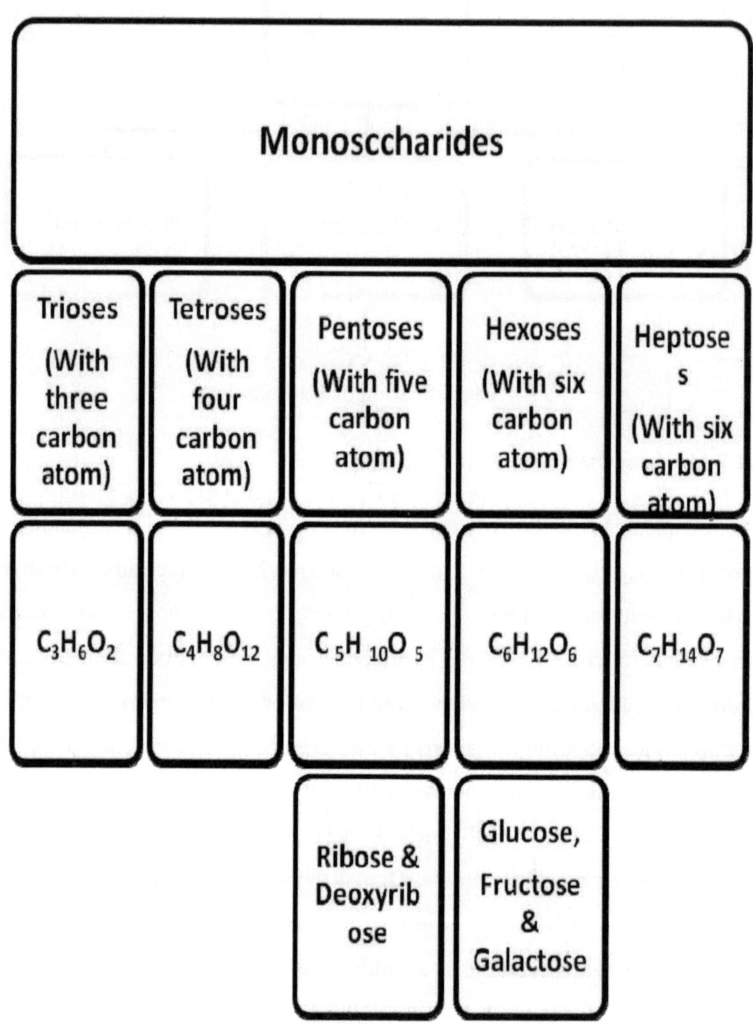

Fig 2. Kinds of Monosaccharides on the basis of Carbon atom

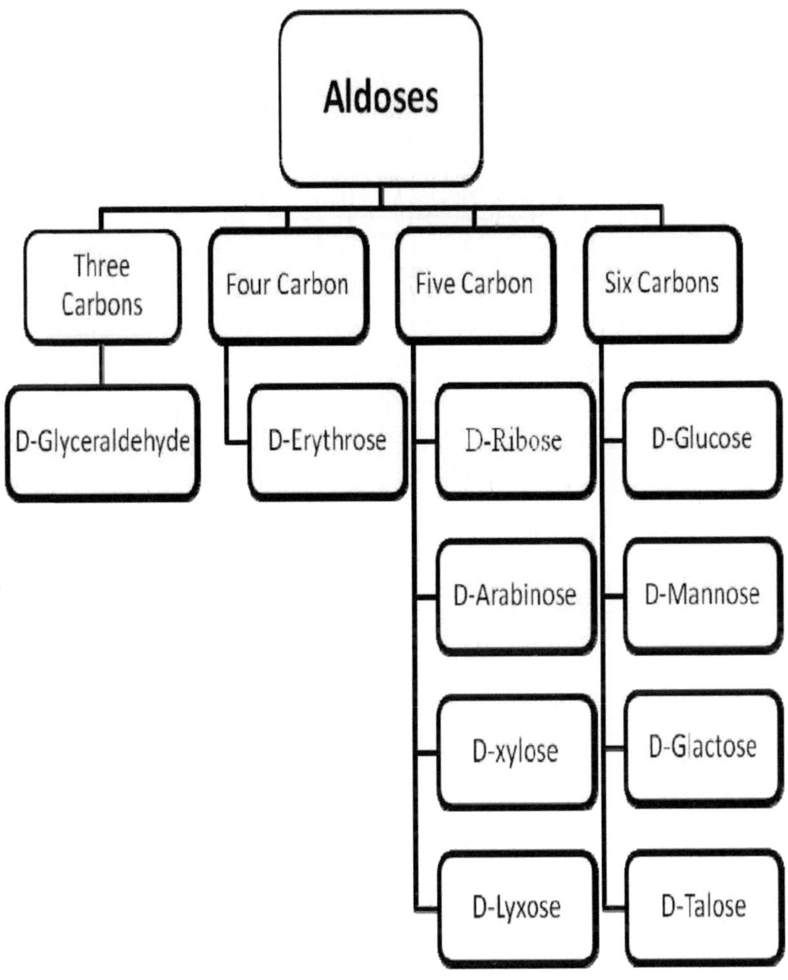

Fig 3.Classification of Carbohydrate on the Basis of Aldose or Ketonic Group and their examples

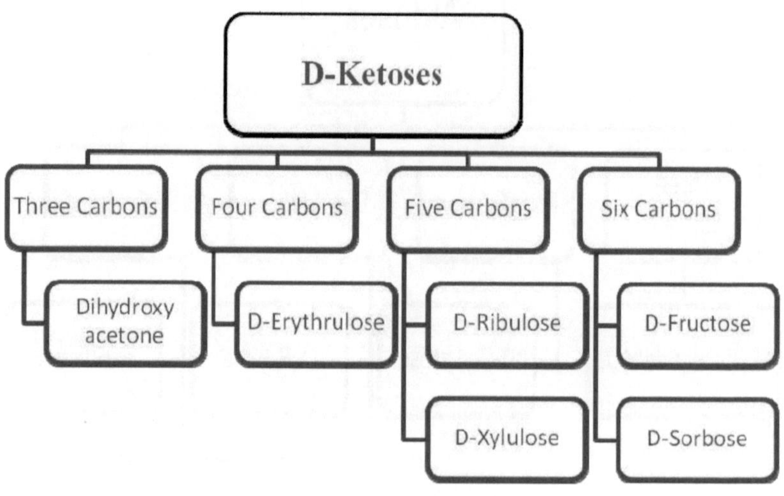

Fig 4. Classification of Carbohydrate on the Basis of Aldose or Ketonic Group and their examples

1.1.3 Oligoscharides:

Oligosachasaccharides consists of short chain of monosaccharide units joined together by characteristic glycosidic linkages. The most abundant are disaccharides, with two monosaccharide units. Typical is sucrose, or cane sugar, which consists of the six carbon sugars D-glucose and D-Fructose joined covalently. All monosaccharide and disaccharides have name ending with the suffix 'ose'. The oligosaccharides are composed of the disaccharides,

8

trisaccharides, and tetrasccharides and are so designated to indicate the number of monosccharides units involved in their structure.

1.1.4 **Disaccharides**: The following tabulation gives the better known disaccharides with their component monosaccharides.

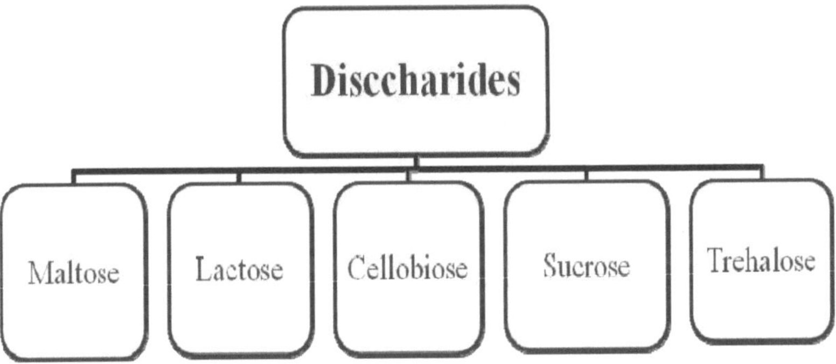

Fig 5.Examples of Disaccharides

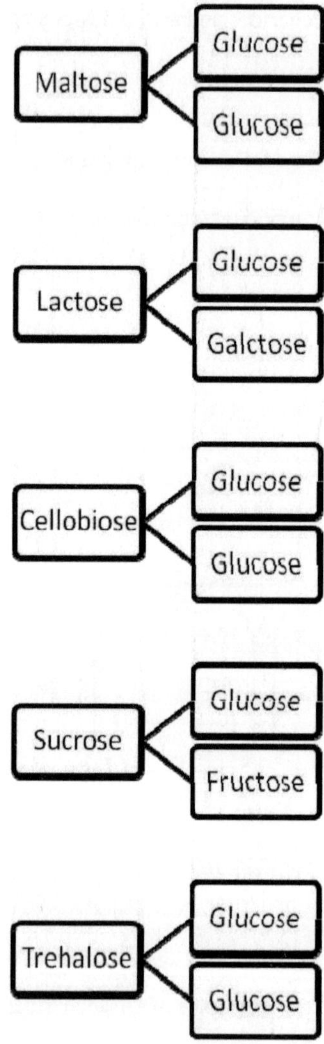

Fig 6. Disaccharides with their component monosaccharide

1.1.5 Trisaccharides:

Several oligosaccharides containing three monosaccharide's units occur in
nature

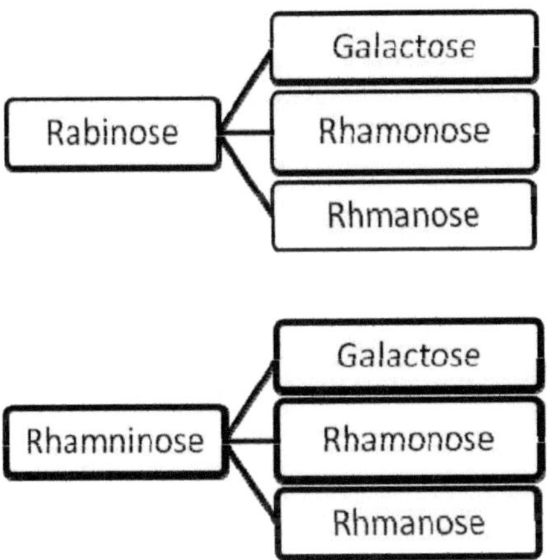

Fig 7. Trisccharides with their component monosaccharide

1.1.6 Polysaccharides:

Polyscharrides consists of long chains having hundreds or thousands of
monosaccharide units. Some polysaccharides, such as cellulose, occur in linear
chains, whereas others, such as glycogen, have branched chains. The most
abundant polysaccharides, starch and cellulose made by plants, consists of
recurring units of D-glucose, but they differ in the type of glycosidic linkage.

1.1.7 General Properties:

1.1.8 Physical Properties:

Monosaccarides are colorless neutral substances having a sweet taste. They do not crystallize very well. On heating they become brown and char without melting. They are readily soluble in water sparingly soluble in alcohol and insoluble in ether.

1.1.9 Chemical Properties:

Aldose and ketoses give all the reactions of aldehyde and ketones respectively, along with those of polyhydroxy alcohols. They also show certain characteristic reactions depending on the presence of the aldehydic or the ketonic group, and the OH groups in the same molecules. The more important reactions of monosaccharides are given below.

1.2.1 **Acetylation:**

Monossacharides undergo acetylation when heated with acetic anhydride and a little anhydrous zinc chloride. Thus, glucose and fructose form penta- acetyl derivatives.

$$C_6H_7O\,(OH)_5 + 5(CH_3CO)_2\,O \longrightarrow C_6H_7O(OOCCH_3)_5 + 5CH_3COOH$$

Glucose   Acetic anhydride   Penta acetyl derivative

1.2.2 Formation of glycosides

Pentoses and hexoses when heated with alcohols in the presence of hydrogen chloride from ethers known as glycosides. Thus glucose and fructose react with methyl alcohol to give methylglucoside and methylfructosides respectively

$$C_6H_{11}O_5OH + HOCH_3 \longrightarrow C_6H_{11}O_5OCH_3 + H_2O$$

Glucose or Fructose   **Methylglucoside**

1.2.3 Formation of salts:

Monosaccharides are very feebly acidic and react with lime to form calcium salts. The glucose and fructose form calcium glucosate (soluble) and calcium fructosate (insoluble) respectively. The calcium glucosate is decomposed by passing carbon dioxide through the solution, thus regenerating glucose. This reaction therefore is employed for the separation of glucose and fructose.

$$C_6H_{12}O_5.OH + HOCa.OH \longrightarrow C_6H_{11}O_5.CaOH + H_2O$$

Glucose   Lime   Calcium glucosate

1.2.4 Reduction:

When monosaccharide is reduced with sodium amalgam in aqueous solution, they take up two H-atom at the CO group and form polyhydroxy alcohols. Thus, glucose yields sorbitol, while fructose yields two isomeric alcohols; sorbitol and mannitol.

1.2.5 Oxidation:

Aldose on oxidation from monocarboxylic acids or dicarboxylic acids containing the same number of carbon atoms, while ketoses oxidize to yields acids with a smaller number of carbon atoms. Thus glucose on mild oxidation with bromide water form gluconic acid.

$$CH_2OH.(CHOH)_4.CHO \xrightarrow{\quad [O]\quad} CH_2OH.(CHOH)_4.COOH$$

Glucose   Gluconic acid

With concentrated nitric acid, however, it is oxidized at both terminal groups and forms saccharic acid

$$CH_2OH.(CHOH)_4.CHO \xrightarrow{\quad [O]\quad} HOOC(CHOH)_4COOH$$

Glucose   Saccharic acid

Fructose remains unattacked by mild oxidizing agents, but when oxidized with concentrated nitric acid. It yield glycolic acid and tartaric acid by cleavage at the CO group

$$CH_2OH.(CHOH)_4.CO.CH_2OH \longrightarrow HOOC(CHOH)_2COOH + HOOC.CH_2OH$$

Fructose   Tartaaric acid

Therefore, glucose and fructose both are reducing sugars. They reduce Fehling's solution and ammonical silver nitrate solution

1.2.6 Formation of Cyanohydrins:

Monosaccharides add a molecules of hydrogen cyanide forming cyanohydrins.

$$CH_2OH.CHOH)_4.CHO + HCN \longrightarrow CH_2OH.(CHOH)_4.CH \begin{matrix} OH \\ \diagup \\ \diagdown \\ OH \end{matrix}$$

Glucose   Glucose Cynohydrin

1.2.7 Formation of Oximes:

Monosaccharide reacts with hydroxylamine to form the oxime by splitting out a molecule of water

$$CH_2OH. (CHOH)_4. CHO + NH_2OH \longrightarrow CH_2OH. (CHOH)_4.CH{:}NOH +$$

$$H_2O$$

Glucose   Glucose oxime

1.2.8 Formation of osazones

Monosaccharides when treated with phenylhydrazine they yield diphenylhydrazone which are known as osazone. According to the mechanism suggested by Fischer, a phenylhydrazone is first produced and then the hydroxyl group adjacent to the original aldehyde or the ketonic group is oxidized to a carbonyl group adjacent to the original aldehyde or the ketonic group is oxidized to a carbonyl group by a second molecule of phenylhydrazine which is reduced to aniline and ammonia. The carbonyl group thus produced now reacts with a third molecule of phenylhydrazine to yield the osazone. It is found that the osazone obtained from glucose and fructose are identical Since the two sugars form the same osazone their structure differ only in respect of two carbon atoms which take part in the formation of osazone. Osazone are yellow crystalline solids having sharp melting points. Their formation is used to characterize the sugar.

Phenylhydrazine ($C_6H_5NHNH_2$) reacts with carbons [#]1 and [#]2 of reducing sugars to form derivatives called osazones. The formation of these distinctive crystalline derivatives is useful for comparing the structures of sugars. Glucose and fructose react as shown below:

Identical osazones are obtained from D-glucose and D-fructose. This demonstrates that carbons [#]3 through [#]6 of D-glucose and D-fructose molecules are identical. The same osazone is also obtained from D-mannose. This indicates that carbons [#]3 through [#]6 of the D-mannose molecule are the same as those of D-glucose and D-fructose molecules. In fact, D-mannose differs from D-glucose only in the configuration of the –H and –OH groups on carbon [#]2.

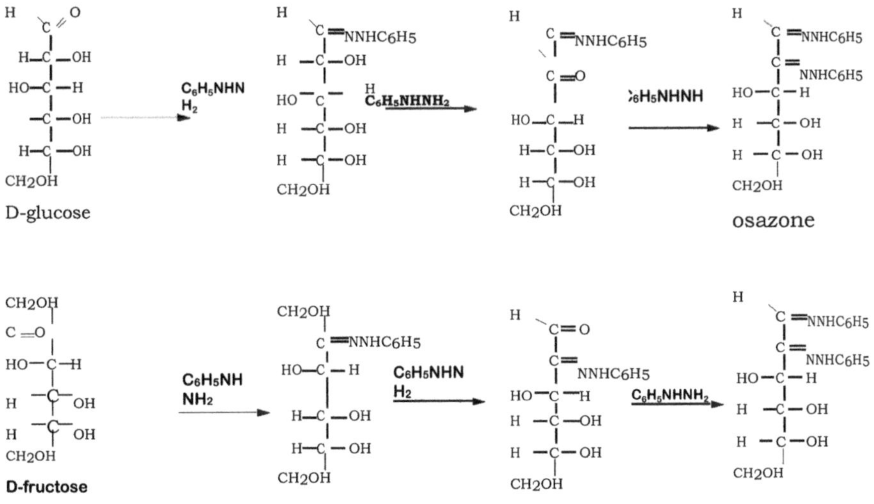

D-glucose $C_6H_5NHNH_2$ → $C_6H_5NHNH_2$ → $C_6H_5NHNH_2$ →   osazone

D-fructose $C_6H_5NHNH_2$ → $C_6H_5NHNH_2$ → $C_6H_5NHNH_2$ →   osazone

1.3.1 Tetroses:

They contain two asymmetric carbon atoms and can exist in four optically active

Forms. They is D-and L-forms of the tetroses, erythrose and threose

1.3.2 Pentoses:

They contain three asymmetric carbon atoms and can exist eight optically active forms. They correspond to the D-and L-form of the four known pentoses; ribose, arabinose, xylose and lyxose.

1.3.3 Hexoses:

They contain four carbon atoms and can exist in sixteen forms which of course, correspond to the D- and L-form of the eight known aldohexoses. These are glucose, mannose, galactose and maltose

Chapter II

EXPERIMENTS No-1: Molisch Test

Object: To perform Molisch test for given sugar solution

Chemical Required:

Concentrated sulphuric acid, α Naphthol, Ethanol

Principle:

Concentrated H_2SO_4 hydrolyse the glycosidic bond to give monosaccharide's which are then dehydrated to form furfural and its derivative like hydroxyl methyl furfural. These furfural then combine with sulphonated α Naphthol to give a purple complex at the junction of two liquid This purple complex is unstable condensation product in the form of ring which disappear on slight disturbance

Procedure:

2-3 drops of α Naphthol solution are added to 2 ml of sugar solution. Very gently 1ml of concentrated H_2SO_4 is added along the side of the test tube.

Observation Table 1:

S.No	Sugar Solution	Observation
1.	1 % Glucose	Purple ring appeared
2	1 % Fructose	Purple ring appeared
3	1 % Galactose	Purple ring appeared
4	1 % Maltose	Purple ring appeared
5	1% Lactose	Purple ring appeared
6	1% Sucrose	Purple ring appeared
7	1% Ribose	Purple ring appeared
8	1% Starch	Purple ring appeared

Result:

All carbohydrate give positive test with Molisch reagent as purple ring appeared in all the test of carbohydrate samples.

Chapter III

EXPERIMENTS NO-2 : Seliwanoff's Test

Object: To perform Seliwanoff's test for given sugar solution to distinguish Ketoses from aldoses

Chemical Required:

Resorcinol, 3N HCl, distilled water

Principle:

This test is used to distinguish ketoses from aldoses. Ketoses are dehydrated more rapidly than aldoses to give a furfural derivative which then condenses with resorcinol to form a red complex. Prolonged heating will hydroyse disaccharides and other monosaccharide which also eventually give colour.

Reagent Preparation:

1. **Preparation of Seliwanoff's reagent (0.05 %, w /v)**

   25 mg of resorcinol was dissolved in little 3N HCl and volume was raised upto 50 ml   by 3N HCl

2. **Preparation of sugar solution**

   1g of sugar was dissolved in100 ml of distilled water

Procedure:

1ml of sugar solution was added to 2 ml of Seliwanoff's reagent and heated in a boiling water bath for one minute.

Observation Table 2:

S.No	Sugar Solution	Observation	Inference
1	Glucose	No colour obtained	Ketose absent
2	Fructose	Red colour (4 min)	Ketose sugar present
3	Galactose	No colour	Ketose absent
4	Maltose	No colour	Ketose absent
5	Sucrose	Red colour (7 min)	Ketose sugar present

Result:

Fructose and Sucrose give positive test with Seliwanoff's reagent. Thus both are seliwanoff ketose sugars. In seliwanoff test red color appears when 1 ml of sample was added to seliwanoff reagent.

Chapter IV

EXPERIMENTS NO-3 Benedicts Test

Object: To perform Benedicts test for given sugar solution to distinguish reducing sugar from non reducing sugar

Chemical Required: Sodium citrate, anhydrous sodium carbonate, copper sulphate, distilled water

Principle:

Benedict test is more convenient and is more stable. In this method sodium citrate functions as chelating agent. Sugar in the presence of strong alkali from enediol and enediol then react with Cu^{++} ions and Cu^{++} ions get converted Cu+. This Cu^{+} ions forms Cu (OH) $_2$ which on heating gives reddish precipitate.

Preparation of Benedict reagent

- 17.5 g of sodium citrate and 10 g of anhydrous sodium carbonate Na_2CO_3 sodium were dissolved in 20 ml of water and volume was raised upto 80 ml by adding distilled water.

- 1.73 g of $CuSO_4.5H_2O$ was dissolved in 5 ml of hot water and volume was raised put 10 ml by hot water and solution was cooled

- Reagent was mixed and volume was raised 100 ml

Preparation of sugar solution

1g of sugar was dissolved in 100 ml of distilled water

Procedure:

0.5 to 1 ml of sugar solution was added to 2 ml of reagent and heated in boiling water Bath.

Observation Table 3:

S.No	Sugar Solution	Observation	Inference
1	Glucose	Reddish orange ppt	Reducing sugar
2	Fructose	Reddish orange ppt	Reducing sugar
3	Galactose	Reddish orange ppt	Reducing sugar
4	Maltose	Reddish orange ppt	Reducing sugar
5	Sucrose	No ppt	Reducing sugar
6	Lactose	Reddish orange ppt	Reducing sugar

Result:

Only those sugar which are reducing in nature gives positives test with Benedicts reagent Sucrose gives negative test, so it is non reducing sugar.

Chapter V

EXPERIMENTS NO-3 Bial's Test

Object: To perform Bial's Test for given sugar solution to distinguish pentoses from hexoses.

Chemical Required: Orcinol, Concentrated hydrochloric acid, $FeCl_3$

Principle:

This test is useful in determination of pentose sugar reaction is due to the formation of furfural and its derivatives in the acidic medium which condense with orcinol in presence of $fe3+$ ions to give blue green complex, which is soluble in butyl alcohol.

1. Preparation of Bial's reagent:

0.3 g of orcinol was dissolved in 20 ml of conc. HCl and 20-30 drops of 10 % $FeCl_3$ solution (2 g of $FeCl_3$ was dissolved in 20 ml of water) was added to it

2. Preparation of sugar solution

1g of sugar was dissolved in100 ml of distilled water

Procedure:

To 2 ml of Bial's reagent 4-5 drops of sugar solution was added. It was heated in boiling water bath.

Observation Table 4:

S.No	Sugar Solution	Observation	Inference
1	Glucose	No colour change	Hexose sugar
2	Fructose	No colour change	Hexose sugar
3	Maltose	No colour change	Hexose sugar
4	Lactose	No colour change	Hexose sugar
5	**Ribose**	**Bial's green complex**	**Pentose sugar**

Result:

Only ribose sugar gives positives test with Bial's reagent thus it is pentose sugar. All other sugar gives negative test with Bial's reagent as they are hxoses.

Chapter VI

EXPERIMENTS NO-5: Mucic Acid Test

Object: To perform Mucic Acid test for given sugar solution to distinguish pentoses from other hexoses

Chemical Required:

Sugar, concentrated HNO_3

Principle:

This test is specific for galactose hence it is distinguished from other monosaccharide by its reaction with conc HNO_3 oxidises carbohydrates to the corresponding saccharic acid. Galactose give characteristic crystals that separate out in conc. HNO_3. Saccharic acid of sugar are soluble except galactose.

Procedure:

50 mg of glucose, fructose and galactose were taken separately in test tube. 1ml of H_2O and 1ml HNO_3 was added to each tube. It was heated in boiling water bath for 1.5 hours it was kept for 24 hours

Observation Table 5:

S.No	Sugar Solution	Observation	Inference
1	Glucose	No crystal was formed	Galactose absent
2	Fructose	No crystal was formed	Galactose absent
3	Galactose	Crystal separates out	**Galactose present**

Result:

Insoluble mucic acid will be formed in galactose but not in glucose and fructose

Chapter VII

EXPERIMENTS NO-6: Barfoed Test

Object: To perform Barfoed test for given sugar solution to distinguish monosaccharides from disaccharides.

Chemical Required:

Sugar, Copper acetate, Glacial acetic acid, distilled water

Principle:

This test is used in distinguishing monosaccharide from reducing disaccharides. Barfoed's reagent is weakly acidic and is only reduced by monosaccharide. Prolonged boiling may hydrolyzed disaccharides to give false positive reaction. Barfoed's reagent is a solution of cupric acetate in dilute acetic acid. When monosaccharide is boiled with this reagent, it is reduced and Cu_2O separates having brick red colour.

Procedure:

2 ml of Barfoed's reagent was taken in test tube and 1 ml of sample solution was added to it. It was kept in boiling water bath. A clean briskly boiling water bath was used for obtaining reliable result.

Observation Table 6:

S.No	Sugar Solution	Observation	Inference
1	Glucose	**Brick red ppt** formed	monosaccharide
2	Fructose	Brick red ppt formed	monosaccharide
3	Galactose	Brick red ppt formed	monosaccharide
4	**Maltose**	**No ppt**	**Disaccharides**

Result:

Glucose Fructose and Galactose are monosaccharide which gives positive test   with Barfoed's reagent.

Chapter VIII

EXPERIMENTS NO-7: Fehling's Test

Object: To perform Fehling's test for given sugar solution to distinguish reducing from non reducing.

Chemical Required:

Sugar, $CuSO_4.5H_2O$, Sodium Potassium Tartrate, potassium hydroxide

Principle:

Fehling's test is a specific and highly sensitive for detection of reducing sugar. Formation of yellow or red ppt of Cu_2O denotes the presence of reducing. Rochelle salt act as chelating reagent in this reaction.

1. Preparation of Fehling's Reagent

- 1.75 g of $CuSO_4.5H_2O$ was dissolved in little distilled water and volume was raised to 25 ml.(**Fehling's solution A**)

- 6 g of KOH and 8.6 g Na-K tartarate was dissolved in water and volume was raised to 25 ml.(**Fehling's solution B**)

- Equal volume of Fehling's solution A and B were mixed

4. Preparation of sugar solution

1g of sugar was dissolved in 100 ml of distilled water and mixed thoroughly

Procedure:

1ml of Fehling's reagent was added to 1ml of sugar solution and heated in water bath.

Observation Table 7:

S.No	Sugar Solution	Observation	Inference
1	Glucose	Red ppt formed	Reducing sugar
2	Fructose	Red ppt formed	Reducing sugar
3	Maltose	Red ppt formed	Reducing sugar
4	Sucrose	No precipitate	Non-Reducing sugar

Result:

Sucrose gives negative test with Fehling's solution, so it is non reducing sugar.

Reaction:

$$CuSO_4 + 2KOH \longrightarrow Cu(OH)_2 + K_2SO_4$$

$$Cu(OH)_2 \longrightarrow CuO + H_2O$$

Chapter IX

EXPERIMENTS NO-8: Iodine Test

Object: To perform iodine test for given sugar solution to distinguish polysaccharides from other sugars reducing from non reducing.

Chemical Required: Iodine, potassium iodide

Principle:

Iodine forms colored adsorption complex with polysaccharide, hence it is a useful, convenient and rapid test for detection of polysaccharides.

Preparation of Reagent:

1. **Preparation of Iodine solution:**

   0.6 g of KI was dissolved in little distilled water and volume was raised to 20 ml (3 % KI solution) 0.02 g of I_2 (0.005 N) was dissolved in it.

2. **Preparation of sugar solution**

   1g of sugar was dissolved in 100 ml of distilled water

Procedure: 4-5 drops of iodine solution was added to 1ml of sugar solution

Observation Table 8:

S.No	Sugar Solution	Observation	Inference
1	Glucose	No colour	Polysaccharides absent
2	Fructose	No colour	Polysaccharides absent
3	Maltose	No colour	Polysaccharides absent
4	Starch	Blue colour	Polysaccharides present

Result:

Only starch gives positive test with iodine solution, so it is a polysaccharide.

Chapter X

EXPERIMENTS NO-9 Osazone test

Object: To perform osazone test for different sugar solution

Chemical Required: Glacial acetic acid, phenyl hydrazine solution, sodium   acetate

Principle:

Monosaccharides when treated with phenylhydrazine they yield diphenylhydrazone which are known as osazone. All monosaccharide or reducing sugars in slightly acidic medium and excess of hydrazine form phenyl osazone compounds. These have crystalline appearance.

According to the mechanism suggested by Fischer a phenylhydrazone is first produced and then the hydroxyl group adjacent to the original aldehyde or the ketonic goup is oxidized to a carbonyl group adjacent to the original aldehyde or ketonic group, is oxidized to a carbonyl group adjacent to the original aldehyde or the ketonic group is oxised to a carbonyl group by a second molecule of phenylhydrazine which is reduced to aniline and ammonia. The carbonyl group thus produced now reacts with a third molecule of phenylhydrazine to yield the osazone.

Procedure:

10 ml of glacial acetic acid was added to 0.5 of phenyl hydrazine hydrochloride and 0.1 g of sodium acetate. 5 ml of test solution was added to it. After gentle shaking it was kept in boiling water bath for 30 minutes

Observation Table 9:

S.No	Sugar Solution	Type and shape of crystal
1	Glucose	Glucosazone (needles or feathery)
2	Galactose	Galactosazone (broad and flat)
3	Lactose	Lactosazone (fine needles grouped in balls)
4	Maltose	Maltosazone, broad needles

Result:

All sugar gives positive osazone test and can be differentiated on the basis of shape of crystals.

Chapter XI

EXPERIMENTS NO-13

Estimation of total carbohydrate (Reducing sugar) by the method of Trevelyan and Harrison (1952)

Principle

The anthrone reaction is the basis of a rapid and convenient method for the determination of hexoses, aldopentoses and hexuronic acids, either free or present in polysaccharides. The blue–green solution shows an absorption maximum at 620 nm, although some carbohydrates may give other color. The reaction is not suitable when protein containing a large amount of tryptophan is present, since a red color is obtained under these conditions. The extinction depends on the compound investigated, but is constant for a particular molecule.

Preparation of Reagent

1. **Sulphuric acid (85 %, v /w)**

   85 ml of sulphuric acid (pre cooled, AR, sp.gr.1.84) was slowly added to 15 ml of DD water and mixed well and cooled in an ice bath.

2 **Anthrone Reagent (0.2 %, w /v)**

   0.2 g of anthrone was dissolved in 100 ml of sulphuric acid (85 % v /v) by stirring it was prepared fresh just before use.

3 Standard Glucose Solution (100 µg /ml)

10 mg of glucose was dissolved in 100 ml of double distilled water. It was stored in refrigerator.

Procedure:

Suitable aliquots (0.1ml to 0.7 ml) were pipetted into a series of boiling tubes and the volume was made upto1.0 ml with double distilled water 6.0 ml of Anthrone reagent (0.2 % w /v) was added slowly with vigorous shaking. The tubes were brought to room temperature and heated for 10 minutes in a boiling water bath. Then after which they were immersed in cold water bath till they attained the room temperature. The intensity of colour was measured at 620 nm. A set of standard glucose (100 µg /ml) and reagent blank was also run simultaneously for the calibration of curve.

Observation Table 10: Estimation of Carbohydrate by Anthrone Method

S. No	Test tube	Glucose ml	Glucose conc µg	Anthrone Reagent ml	Double distilled Water(ml)		A620nm
1	0	0.0 ml	00 ug	6	1 ml	Heated	00
2	1	0.1	1	6	0.9	In water	0.07
3	2	0.2	2	6	0.8	bath for	0.14
4	3	0.3	3	6	0.7	10 min	0.20
5	4	0.4	4	6	0.6	&	0.28
6	5	0.5	5	6	0.5	cool at	0.35
7	6	0.6	6	6	0.4	room temp	0.48

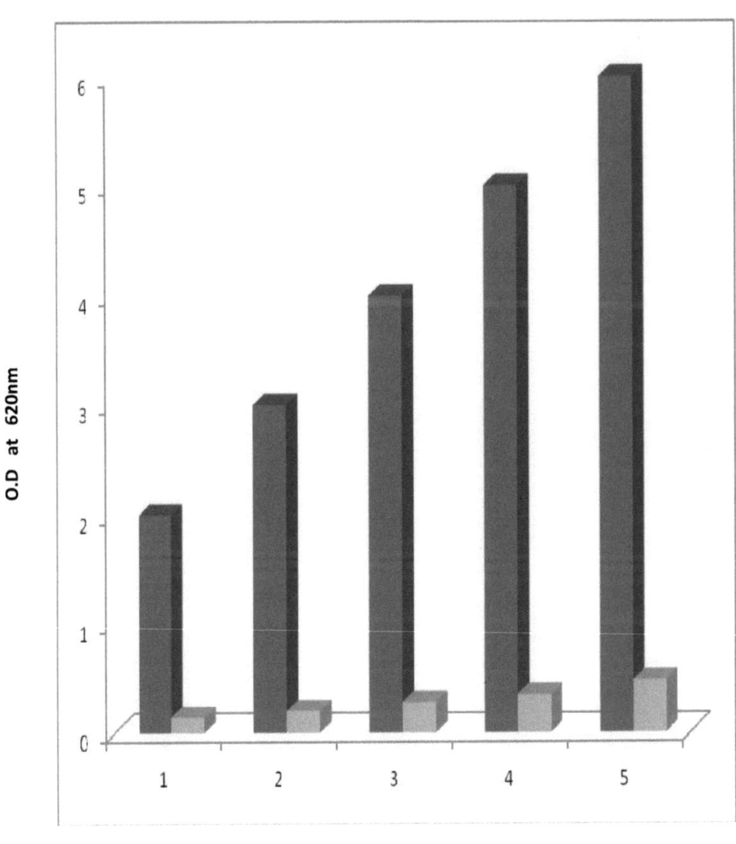

Fig 8. Estimation of Carbohydrate by Anthrone Method

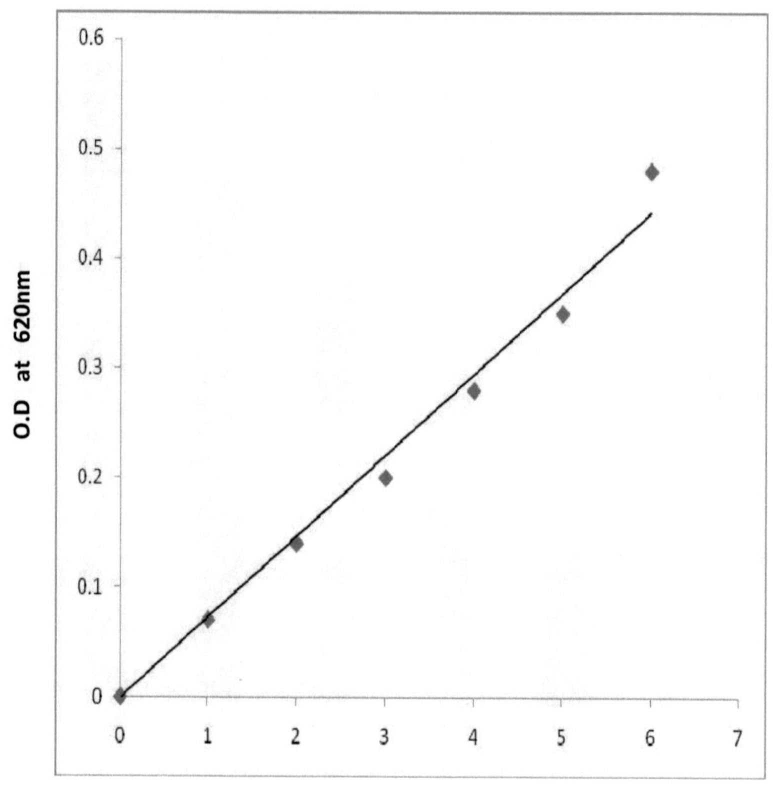

Fig 9. Estimation of Carbohydrate by Anthrone Method